IMAGES
of America

MARINE CORPS
AIR STATION
EL TORO

IMAGES of America
MARINE CORPS AIR STATION EL TORO

Thomas O'Hara

ARCADIA

Copyright © 1999 by Thomas O'Hara.
ISBN 0-7385-0186-7

Published by Arcadia Publishing,
an imprint of Tempus Publishing, Inc.
2 Cumberland Street
Charleston, SC 29401

Printed in Great Britain.

Library of Congress Catalog Card Number: 99-63926

For all general information contact Arcadia Publishing at:
Telephone 843-853-2070
Fax 843-853-0044
E-Mail arcadia@charleston.net

For customer service and orders:
Toll-Free 1-888-313-BOOK

Visit us on the internet at http://www.arcadiaimages.com

MCAS EL TORO COMMISSIONING CEREMONY, MARCH 17, 1943. (Photograph from Flying Leatherneck Aviation Museum (FLAM) Library.)

CONTENTS

Introduction		7
1.	Early El Toro through World War II	9
2.	Late 1940s: Air FMFPac & 1st MAW	23
3.	A New Age in Aviation and Korea	49
4.	3d MAW Arrives at El Toro	59
5.	A Few Famous Marines	67
6.	The Sixties and Seventies	73
7.	Base Renewal	83
8.	1990s, Desert Storm & BRAC	97
9.	MCAS Tustin (LTA)	109

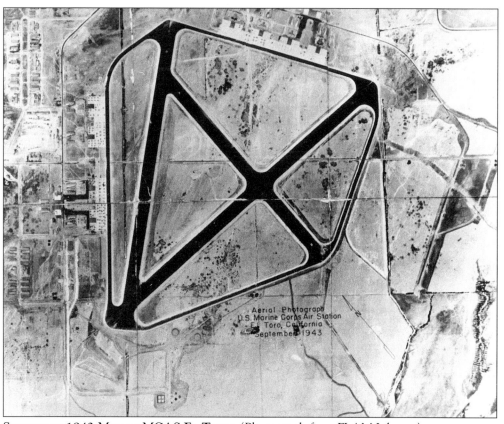

SEPTEMBER 1943 MAP OF MCAS EL TORO. (Photograph from FLAM Library.)

INTRODUCTION

For those who live there, southern California is a paradise. In addition to the great weather, proximity to many natural and man-made attractions, and strong economic base, it is rich in its historical heritage. The area, after being invaded by the Spanish, has been ruled under the flags of four nations: Spain, Mexico, California (prior to statehood), and the United States. Don Gaspar de Portolal first claimed the area for Spain on St. Anne's Day, July 26, 1769. The first Spanish settlers holding title to the land were Jose Spulveda, who received a grant to Rancho San Joaquin from Mexico in 1836, and Theodosia Yorba, who acquired Rancho Lomas de Santiago in 1840.

In 1860, San Francisco merchant James Irvine Sr. acquired the 107,000-acre property from its Spanish owners in payment for a debt. James Irvine Sr. experimented with farming the land, which previously had been used only for grazing cattle and horses. In 1906, James Irvine's son moved to Orange County after surviving the San Francisco earthquake to manage the Irvine Ranch. He converted much of the prime land from grazing to agricultural pursuits. Various grain crops grew in the area until that time when an experimental field of lima beans and black beans became so successful that eventually 20,000 acres were devoted to beans alone.

The military first considered Orange County as a site for an air station in 1928. Chief of Naval Aviation Rear Admiral W.A. Moffett landed at the Eddie Martin Airport (now Orange County airport) and spoke with James Irvine, owner of the famous Irvine Ranch, about purchasing two large lima bean fields for use as military airfields. One of the fields was located near the mouth of Canada Del Toro (Canyon of the Bull) in what is now MCAS El Toro, and the other was located in what is now MCAS Tustin, California. Mr. Irvine did not meet Admiral Moffet's first encounter with much enthusiasm. Mr. Irvine was not excited about losing his two most productive bean fields, and luckily the navy chose a site in northern California. However, shortly after the bombing at Pearl Harbor, Hawaii, the Marine Corps came calling a second time.

On May 5, 1942, Lt. Col. William J. Fox received a message that designated him as the navy representative to the site selection committee to make immediate surveys of southern California to determine the best available sites for Marine Corps air bases. He was told that first consideration should be given to a site that would relocate the five Marine Carrier Groups serving at NAVAIRSTA San Diego. Marine Corps officials favored fives sites—Santa Barbara, El Centro, Mojave, Miramar, and the bean fields at El Toro.

On Admiral Moffet's advice, Lieutenant Colonel Fox came to southern California to see the bean fields. James Irvine offered Lieutenant Colonel Fox other sites on his ranch for $1 a year, including the Orange County Airport site, attempting to preserve the bean field crop. The Marines decided they wanted the bean fields for their airfield. In early 1942, the Marines paid James Irvine $100,000 for 4,000 acres, including 1,600 acres designated for a lighter-than-air (blimp) base.

The bean field that became MCAS El Toro appeared to be perfect for Marine Corps requirements. The field sat in a quiet valley at the base of the Saddleback Mountains. Only a few miles from the Pacific Ocean, it was fog free most of the year, and the main Santa Fe Railroad line ran adjacent to the site's west border. The bean field also perched just north of Camp Pendleton, which served as the Marines Infantry Training Base and bombing range. El Toro was also close to San Clemente Island, approximately 60 miles off the coast, which also served as a bombing range. The seaport of Long Beach just north of El Toro was used extensively during and since World War II (WW II) to ferry troops to overseas war zones, to disaster sites needing humanitarian assistance, and as a debarkation point for Marines serving on ships around the world.

Government officials and businessmen in Orange County, unlike James Irvine, were extremely happy the Marines had landed. The new air base would bring thousands of civilians and military people into the area, as well as jobs and money. These politicians knew the military base had given birth to what would become one of the wealthiest counties in the nation.

On September 23, 1942, Col. Theodore Millard arrived in southern California to take charge of base construction. While pondering the enormous task before him, he received a message that the first 30 Marines to be assigned to MCAS El Toro had arrived at the nearby train depot. No military transportation existed at the time, and no barracks had been built to house the Marines. For the next several hours Colonel Millard drove his private vehicle back and forth, retrieving the Marines for duty. The 30 Marines came prepared to pitch tents and live "in the rough" as they usually do, but Colonel Millard borrowed a bunkhouse on the Irvine Ranch, which, by comparison, was a substantial upgrade. Gy. Sgt. Howard W. Langdon Sr. recalls being one of the original 30 enlisted men to arrive at El Toro, "back in 42." His fondest memories involve living in the Irvine Ranch bunkhouses and eating breakfast each morning at the ranch chuck wagon.

Construction on the air station began on August 3, 1942, less than a year after the bombing at Pearl Harbor. Within four months, construction crews finished the runways, taxiways, and run-up areas. A month later, in January 1943, the first hangars, barracks, and bachelor officer quarters were completed, and the finishing touches turned a bean field into an air station. Five asphalt runways and over 27 miles of paved highways were laid in five months.

The effort, although incredibly successful, did not proceed without problems. In the very early days, the few wooden buildings all had very leaky roofs. At one point, heavy rains turned the buildings into wash tubs and the freshly turned dirt into a giant mud bowl; there were no laundry facilities to clean muddy uniforms. One of the major disappointments occurred at the new commissary building. While the roofers were hot-tarring the roof, the building caught fire and burned to the ground. However, by December 1942, planes were taking off and landing. When the skies were clear, Marines and construction workers alike watched the mock dog fights overhead as pilots trained for combat in the Pacific. One of the Marine Corps' finest installations sprung from a bean field in sunny southern California.

One
Early El Toro through World War II

Marine Base Defense Group 41 (MBDAG-41) was organized at El Toro on January 1, 1943. Originally, MBDAG-41 was created to administer and supervise training and activities of aircraft squadrons for combat duty in the Pacific. Consisting of only a Headquarters Squadron, a Service and Maintenance Squadron, and one Fighter Squadron on January 1, MBDAG-41 eventually trained 20 Fighter, Dive Bomber, and Torpedo Squadrons for combat duty by the end of the war. Headquarters Squadron 41 was probably the largest headquarters in Marine Corps history at peak strength, with approximately 2,000 personnel.

During the same month, flight operations at El Toro accelerated into a high tempo. The first squadron assigned to El Toro was VMF-113, flying SNJ trainers and F4F Wildcats. Combat-tested flying squadrons VMF-212, VMF-223, VMF-224, and VMSB-232 arrived from the Pacific to re-fit and reorganize. Squadrons flying F4U Corsairs, Douglas SBDs, and Curtis SB2C dive bombers arrived from NAS North Island, MCAD Miramar, and from combat duty in the Pacific. In addition, B-24 Bombers from the consolidated plant in San Diego lumbered down the runways en route to San Clemente Island for bombing practice. On January 10, 1943, these squadrons fell under the command of the newly created Marine Aircraft Group 23. During WW II, El Toro was used primarily as a training base for pilots, aircrews, and ground personnel, but was also a major debarkation station for personnel being transferred to overseas duty.

> War is an ugly thing, but not the ugliest of things,
> the decayed and degraded state of moral and patriotic feeling
> which thinks that nothing is worth fighting and dying for is much worse—
> a man who has nothing for which he is willing to fight;
> nothing he cares about more than his own personal safety;
> is a miserable creature who has no chance of being free
> unless made and kept so by the exertions of men
> far better and braver than himself . . .
>
> by John Stuart Mill

EARLY EL TORO. This early aerial photograph shows El Toro's Main Gate to the Control Tower. Notice the absence of trees and shrubs on the base.

COMMISSIONING CEREMONY. The El Toro Commissioning Ceremony on St. Patricks Day, March 17, 1943, had the pomp and ceremony expected of an occasion of this magnitude. A military parade followed the official flag raising and fly-over by SBD Dauntless Dive Bombers and F4F Wildcats.

THE SECOND C.O. On June 7, 1943, Col. William J. Fox, a Guadalcanal veteran, who supervised the construction of Henderson field and served on the initial El Toro site selection committee, became the second commanding officer of El Toro.

AWARDS CEREMONY ON JUNE 19, 1943. Lt. Col. "Indian Joe" Bauer received the Medal of Honor posthumously during this ceremony. After an intense air battle over Guadalcanal, he was shot down just short of landing at Henderson Field.

EL TORO'S MAIN GATE. Traffic during the war was light, for few people owned automobiles. Gasoline rationing limited travel as well.

WILDCATS READY FOR FLIGHT. These Wildcats betray the otherwise peaceful serenity going on around them.

TBM AVENGER. The TBM Avenger is based on the TBF design. The TBF first flew in 1941. Nearly 10,000 were built by Grumman and General Motors. The TBF/TBMs were used primarily as torpedo bombers.

WARBIRDS. A F4U-5N Corsair and a F6F-5N Hellcat from VMF(N)-513 can be seen in this photograph.

CORSAIRS. F4U-4s on the El Toro flight line prepare for takeoff. (Note the F7Fs in the background.) The Corsair is often referred to as the Marine Corps signature aircraft from WW II. The Corsair ended WW II with a 12-1 kill ratio.

WILDCAT. This F4F is flying over southern California. The F4F was the primary fighter for the USMC/USN until 1943. Built by Grumman and Eastern aircraft companies, it was one of the slowest fighters. The Wildcat was the first American fighter to sink a Japanese ship in the Pacific.

DOUGLAS SBD FLYING OVER SOUTHERN CALIFORNIA. During the Battle of Midway, First Lieutenant Daniel Iverson made it back to Midway Island with 259 bullet holes in his SBD and his throat microphone shot away. The SBD was a sturdy aircraft that saw action throughout the Pacific during WW II.

CHANCE VOUGHT F4U-1 CORSAIRS. Corsairs were also built by Goodyear and Brewster. The folded wings reduce the aircraft's footprint, allowing many more aircraft to be flown from aircraft carriers.

The SNJ. Nearly 16,000 SNJs were built during WW II for the military. The navy and Marine Corps flew them as training aircraft. The Pratt & Whitney engine had 550 HP.

Hellcats. This F6F-5N is on El Toro's flight line. The F6F ended WW II with a 19-1 kill ratio.

SOUTH PACIFIC. This south Pacific setting seems tranquil, but the successes indicated on the side of aircraft #122 indicate otherwise.

THE PBS. The PBS or B-25 played an important role in the pacific theater. It was a Bomber, it could provide close air support, and one model had an artillery piece on the nose.

F4U. This F4U Corsair from UMF-214 made a gear-up landing at Bauganville in 1943.

EMERGENCY LANDING. The pilot flying this Corsair made an emergency, gear-up landing in the desert. The pilot walked away with no injuries.

A Flight of Corsairs Looking for Trouble.

MARINES ALWAYS FIND WAYS TO ENTERTAIN THEMSELVES. This sign facetiously depicts the absence of amenities on this worn torn island.

FORWARD AIR BASE IN THE SOUTH PACIFIC. F4Us and PBJs stand ready while a transport departs in defiance of hostile skies.

THE SCAT (SOUTH PACIFIC COMBAT AIR TRANSPORT) TERMINAL. This was the only consistent air transportation between Guadalcanal and the outside world. MAG-25 saved many lives by evacuating 2,879 casualties off Guadalcanal. This number was three times more than evacuated by ship.

THE BLACKSHEEP. The famous 214 Blacksheep logo can be seen on a Corsair engine cowling.

Two

Late 1940s

Air FMFPAC & 1st MAW

At the end of WW II, El Toro found itself on a short list of bases to be closed. However, in 1946, Headquarters Marine Corps placed El Toro on its list of seven installations to be maintained in active status. Shortly thereafter, the commander of Marine Air Base West Coast, Major General Louis E. Woods, moved his headquarters to El Toro from MCAD Miramar, and the last remnants of the Marine presence at Miramar followed.

Aircraft, Fleet Marine Force, Pacific was originally Marine Aircraft Wing, Pacific, and was organized at the Naval Air Station San Diego on August 15, 1942. Its mission included the organization, administration, and assignment of personnel and supplies for the First and Second Aircraft Wings, the Fourth Marine Base Defense Wing, and the Headquarters Squadron and a service group. On September 16, 1944, MAWPac was re-designated Aircraft, Fleet Marine Force, Pacific. On October 21, 1944, two additional commands were added to Air FMFPac. Provisional Air Support Command headquartered at Ewa, Hawaii, and Marine Carrier Groups at Santa Barbara were organized and placed under command of Air FMFPac. As the parent organization of all Marine aircraft in the Pacific, Air FMFPac administered, organized, deployed, and supplied every Marine aviation unit that fought the Japanese from Guadalcanal to Okinawa.

From the end of WW II until May 1949, Air FMFPac was based at Ewa, Hawaii. On May 3, 1949, Air FMFPac returned to the United States and consolidated with the First Marine Aircraft Wing under the command of Major General Louis E. Woods. Air FMFPac declined to skeleton proportions, but in little more than a year, the Korean War would require revitalization of this command unit. Air FMFPac returned to MCAS El Toro upon conclusion of the Korean War, but relocated to Hawaii after the 3d MAW arrived at El Toro.

In 1949, the 1st Marine Air Wing arrived from Tientsing, China, bringing the huge Taoist bell, which has been an El Toro landmark ever since. The bell was cast in HoChen, China, north of the Yellow River in 1628, where it hung for hundreds of years. The Japanese took the bell to Tientsing to salvage the metal during the war. The 1st Marine Air Wing arrived in time to save the town and the bell. The bell has hung in front of the El Toro headquarters building since 1949. When the Korean War erupted in 1950, the 1st Marine Air Wing re-deployed to the Pacific theater, where it has stayed.

The first Marine jet squadron to fly in combat (VMF-311) in Korea came from El Toro, departing in November 1950 with F9F-2B Panthers. Post-WW II defense cutbacks did little to prepare the nation for the Korean War. The existence of bases like El Toro expedited the military's ability to overcome these deficiencies. Upon mobilization, and calling upon the military reserves to fill the holes in the nation's defense capability, El Toro served as a major reporting, processing, training, and departure point for thousands of reserves called to duty to serve in Korea.

THE EARLY AIR FMFPAC LOGO.

THE EARLY 1ST MAW LOGO.

PUBLIC TRANSPORTATION. After WW II, the 1st MAW remained in China until 1949. In this photo a few marines take advantage of the local public transportation system.

THE TAOIST BELL BROUGHT TO EL TORO BY THE 1ST MAW. This bell is a symbol of the countless ways the Marine Corps has served the nation and the far away "clime and places" where they serve.

SNOW ON EL TORO. An extremely unusual snowstorm blanketed El Toro in 1949.

FLIGHT OPERATIONS PARALYZED. The snow covered the aircraft, runways, and nearby hills paralyzing flight operations. Since no snow removal or de-icing equipment was available, the El Toro Marine Corps was grounded until the snow melted.

THE MARAUDER. The Martin JM-1 Marauder in the hangar was used briefly by the Marine Corps.

THE FIRST HELICOPTER LANDING. This photograph shows the first helicopter landing at MCAS El Toro during the late 1940s. The Marine Corps field tested several helicopters after WW II and formed the first helicopter assault squadron in 1951.

POST FLIGHT. A VMF 232 Corsair gasses up after a morning flight. The number 13 is awarded by some superstitious squadrons. During the Korean War, one squadron put the number 12 7/8 on a replacement aircraft after its two predecessors (both number 13) were lost in combat action.

THE SHOOTING STAR. TO-1s are on the El Toro flight line. The TO was the first mass production jet in the Marine Corps.

THE FLYING PEONS. Enlisted pilots were a special breed in the Marine Corps. They called themselves "Flying Peons." For over 50 years, Marines wearing chevrons piloted all types of Marine Corps aircraft. The first two enlisted pilots received formal training in 1916 and flew combat missions in France during WW I. Stories of valor and heroism abound in this unique group of Flying Devil Dogs.

CONTROL TOWER PERSONNEL. The tower crew monitors flight operations, controls airfield airspace, and provides pilots with takeoff and landing data.

Painting aircraft. Notice the 1940s compressor and spraygun unit used in this photograph.

Boresighting a Wildcat. Ordinance personnel boresight this aircraft the old-fashioned way.

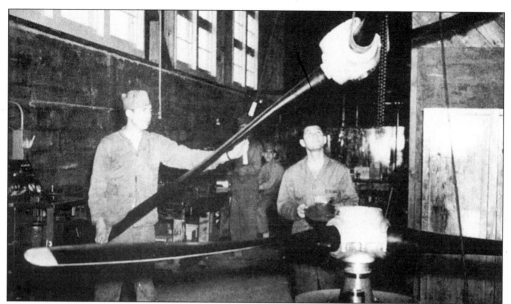

THE PROP SHOP. The propeller shop repairs and realigns propellers. Many of these large propellers spun precariously close to the ground upon takeoff and landing. Pilots had to be very careful to avoid hitting the ground with their props.

BREAK TIME. Mechanics take a break in this photograph. These unsung heroes work long and hard to keep aircraft flying; they never seem to get the credit they deserve. For every hour of flight time, many hours of maintenance must be performed. After combat missions, that ratio is often quadrupled.

ORIGINAL "FLYING BULL" LOGO. This is a copy of the original "Flying Bull" logo prepared by Walt Disney and given to the Officers of the Marine Corps Air Station at El Toro, California.

Large Transport Aircraft Line-up in front of the New MCAS El Toro Control Tower.

THE FLYING BULL. This version of the El Toro Flying Bull logo appeared in a 1944 National Geographic magazine. Walt Disney was asked to design an insignia for the new air station resulting in the famous "Flying Bull," which existed in nearly original form at the time of base closure.

THE WOMAN RESERVES TAKING OVER. Woman Reserves pledge the oath. On November 14, 1943, four Woman Reserve Officers and 96 Woman Reserve enlistees arrived at El Toro to operate Link trainers, teach recognition, teach gunnery, and perform other duties, which released male Marines for combat duty. The Woman Reserves were later designated Woman Marines and continue to serve proudly to this day.

WR Weather Forecasters. Two WRs are preparing to release a weather balloon.

WRs Working on a PBJ. During WW II, women performed almost every job previously performed by males.

WRs Working in the Motor Pool.

Laundry. The laundry detail has always been one of the largest and most important duties in the Marine Corps.

FLIGHT SCHEDULE. A moment in time is captured by this schedule.

WRs. WRs packed parachutes and ran the paraloft. Few positions are held in higher esteem by pilots and aircrew than the parachute packer.

A YOUNG BOB HOPE. Bob Hope loved the Marine Corps as much as the Corps loved him. Surrounded by a Woman Marine Fire Team, Jean Irving (upper left) catches his eye.

AN EARLY CRASH CREW VEHICLE.

A PILOT'S BEST FRIEND. The Fire Department and Crash Crew stayed busy during the early years at El Toro. Unfortunately, the aircraft were difficult to fly and the training was sometimes hurried, which resulted in an abnormally high number of accidents. This landing obviously created more work for the propeller shop.

CRASH CREW VEHICLE.

CRASH CREW VEHICLE.

HAPPINESS IS . . . Marines are always happy when they are well fed.

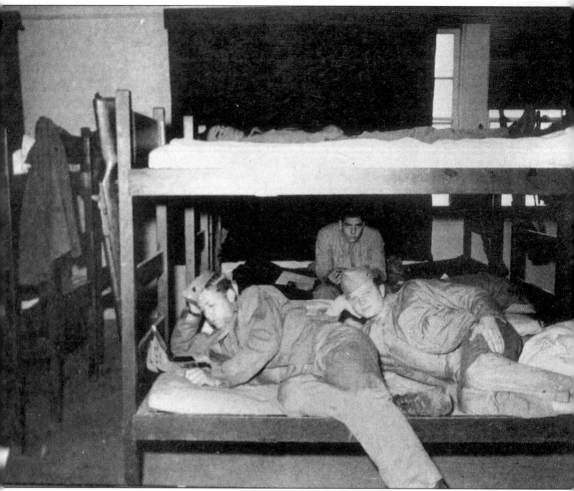

THE OLD CORPS. Old Corps Marines remember the community living offered by squad bays and bunk beds. In this picture, the Marine's rifle is neatly hung "sling arms" on the end of his bunk.

FREE TIME. Outside each of the barracks, the Corps graciously provided facilities to clean rifles, 782 Gear, and laundry during a Marine's free time. These daily opportunities allowed senior Marines time to counsel junior Marines about the world "according to the sergeant mayor."

FREE TIME. Billiards has always been a favorite recreation enjoyed by Marines everywhere.

A MILITARY POLICEMAN AND HIS MASCOT.

MARINES HAVING FUN. For young men entering the Marine Corps, there is an abundance of off-duty activity available. In 1943, El Toro was a remote location, and public mass transportation did not exist. Disneyland would not be built for another decade. The base provided a variety of sports and social functions to keep morale high.

Station Gymnasium.

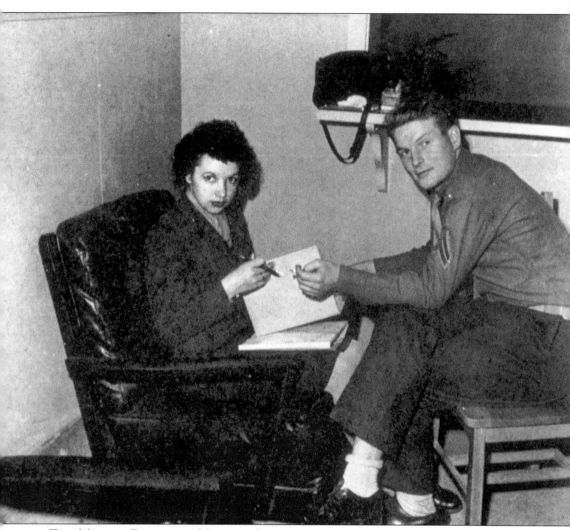

Two Marines Discussing Military Strategy.

Three
A New Age in Aviation and Korea

The first prototype jet aircraft had been flying by the end of WW II. After the war, defense cutbacks slowed the development of jet aircraft programs. Jet aircraft were especially slow coming to the Navy and Marine Corps because additional research and development was required to produce aircraft that could operate aboard aircraft carriers. VMF-311 was the first squadron at El Toro to fly a jet aircraft named the *Shooting Star*. The Lockheed F-80C *Shooting Star*, re-designated the TO-1 by the Navy, became the military's first production jet fighter in 1948. In March 1950, VMF-311 transitioned to the F9F Panther prior to departing for Korea.

This period in aircraft development represented one of the great leaps in design and technology. During this period, jet aircraft designs proliferated, eventually providing the Marine Corps with the Banshee, the Panther, the Skyknight, the Cougar, and the Fury. The advent of jets did not spell immediate doom for propeller aircraft. Several propeller-driven fighters remained in service until the mid-1950s, while the Douglas Skyraider was used throughout the Vietnam War; some of the old transports are still flying today for firefighting companies and at FBOs around the world. In 1950, North Korean troops stormed across the 38th parallel into South Korea. This second conflict in the Pacific in less than a decade confirmed the Marine Corps' need for a permanent west coast air base to ensure combat-ready pilots were available and as a debarkation point for Marines going overseas. On January 15, 1951, HMM-161 was activated at MCAS El Toro. HMM-161 is the first and oldest tactical helicopter squadron in the Marine Corps. In September 1951, HMM-161 arrived in Korea to become the first helicopter squadron to prove itself in combat. The Korea War lasted three long, cold, and bloody years. Jet aircraft were christened under fire, and a new age in aviation was born.

High Flight by John Gillespie Magee Jr.

Oh, I have slipped the surly bonds of earth
and danced the skies on laughter-silvered
  wings;
Sunward I've climbed, and joined the
  tumbling mirth.
Of sun-split clouds—and done a hundred
  things
You have not dreamed of—wheeled and
  soared and swung
high in the sunlit silence. Hov'ring there,

I've chased the shouting wind along, and
  flung
my eager craft through footless halls of air
Up, up the long delirious, burning blue
I've topped the windswept heights with easy
  grace
Where never lark, or even eagle flew.
And, while with silent, lifting mind I've trod
The high untrespassed sanctity of space
Put out my hand, and touched the face of God

THE DEATH RATTLERS. VMF-323 were also known as the Death Rattlers. Here, VMF-323 are embarked aboard an aircraft carrier for duty in Korea.

VMF-212: THE DEVIL CATS. Here, VMF-212 poses in full regalia in front of their Corsairs. The F4U Corsair was in production longer than any other U.S. WW II fighter. Its production time spanned from first flight in May 1940 until the last Corsair was delivered in December 1952.

K-1. This expeditionary airfield in Korea was known as K-1.

Cold Snowy Korea. This Corsair stands ready for takeoff in a snowy Korean Bunker.

THE DOUGLAS F3D SKYKNIGHT. Conceived in 1946 as an all-weather fighter, it made its first flight in March of 1948. The F3D was modified to conduct electronic counter measures missions, and as missile platforms. A VFM(N)-513 F3D became the first jet to shoot down an enemy aircraft at night. The F3D destroyed more enemy aircraft in Korea than any other Navy or Marine Corps aircraft. The F3D's long service life allowed it to fly well into the Vietnam conflict until it was replaced by the Grumman EA-6A.

A FLIGHT OF F3DS. This flight of skynights is loaded with bombs en route to enemy targets.

F3Ds on the Flight Line.

THE GRUMMAN F7F TIGERCAT. The first F7Fs were delivered to the Marine Corps in 1944 and served throughout the Korean War.

THE DOUGLAS AD SKYRAIDER. Known by some as the "Flying Dump Truck," the Skyraider could carry more bombs than the four-engine B-17 bomber.

THE F2H-4 BANSHEE. The prototype for the Banshee was the FH-1 Phantom, which first flew in 1947 while the advanced model F2H-2 first flew in 1949. The -3 and -4 models stayed in service in the Marine Corps until 1959.

THE F9F PANTHER. This aircraft achieved many firsts in flight. It was Grumman's first jet fighter, it was the first Navy jet to see combat, and it was the first Navy jet to shoot down an enemy jet. The first production Panther flew in 1949.

THE GRUMMAN F9F COUGAR. The Cougar was a swept wing version of the Panther. The first F9F-6 Cougar flew in 1951. The last F9F-8Ps were delivered in 1957. This photograph shows a flight of four Cougars over Newport Beach, California.

A Grumman Cougar over MCAS El Toro. By this time the airfield had been upgraded with two sets of parallel runways and additional taxiways.

The North American FJ-3 Fury. The FJ-3 was a revised FJ-2 with larger engines. The FJ-2 was a version of the F-86 Sabre jet that was modified for naval operations.

THE FIRST HELO ASSAULT SQUADRON. The Sikorsky HRS or H-19 Helicopter was flown by HMR-161. This helicopter was the Marine Corps first assault helicopter; the HMR-161 comprised the first squadron and was deployed to Korea in 1951 shortly after formation.

OBSERVATION AND MEDICAL EVACUATION HELICOPTER. Helicopter appearances at MCAS El Toro became more frequent when the Marine Corps assumed control of Marine Corps Air Facility Santa Ana, California. This observation helicopter is equipped with an evacuation litter.

THE KAMAN HOK HUSKIE HELICOPTER. This helicopter was used as the MCAS El Toro search and rescue helicopter and for basic utility functions.

THE PIASECI HUP RETRIEVER. This helicopter was designed for the navy for shipboard operations. The Marine Corps only owned 13 HUPs and used them primarily for search and rescue.

Four
3D MAW ARRIVES AT EL TORO

In 1955, the Third Marine Aircraft Wing arrived at El Toro from Opa Lacka (near Miami), Florida. The 3d MAW, activated on the Marine Corps Birthday, November 10, 1942, at MCAS Cherry Point, North Carolina, was created to train pilots and organize squadrons for combat duty in WW II. During WW II, the 3d MAW deployed 10 Air Groups, 11 Early Warning Squadrons, and 37 Tactical Fighter Squadrons. In April 1944, the 3d MAW relocated to MCAS Ewa, Hawaii, to support the island fighting against the Japanese. Upon the conclusion of the war, the 3d MAW was deactivated. As the Korean War expanded, the 3d MAW was reactivated in 1952 at Cherry Point and relocated to Miami, Florida. In 1955, the 3d MAW again relocated to MCAS El Toro.

El Toro's early 3d MAW consisted of fighter, attack, photograph reconnaissance, observation, refueling, and bomber jet aircraft, as well as, attack, utility, and transport helicopters, and C-130 transport planes. In addition, training schools and courses included aircraft instrument school, maintenance schools, survival, ejection seat, low pressure and water survival schools, aircraft simulator training, Nuclear, Biological, and Chemical Warfare School, jet transition squadron schools, special weapons and tactical atomic weapons school, and the Staff Noncommissioned Officers School among others.

> I am convinced there is no smarter, handier or more adaptable body of troops in the world than the U.S. Marines . . . Always spick and span, ready at an instant's notice for duty, the nation owes them a great debt.
> —Winston Churchill

THE THIRD MARINE AIRWING LOGO.

THE OLD THIRD MARINE AIRWING HEADQUARTERS BUILDING AT MCAS EL TORO.

THE VOUGHT F8U CRUSADER. The F8 Crusader was the first carrier-based fighter that could fly in excess of 1,000 mph. The Crusader first flew in 1955 and was retired by the Marine Corps in 1975.

THE NORTH AMERICAN SAVAGE REFUELING TWO FURIES. Aerial refueling can be very difficult and sometimes impassible. Prop wash from the refueler, turbulence, thermals, and pilot skill all play important roles in success or failure.

AN AERIAL WEST VIEW OF MCAS EL TORO. On a clear day, Santa Catalina Island can be seen 26 miles off the coast. At top left, the El Toro golf course is located behind the MAG-13 area. At center left the MAG-11 area is located behind the central tower. On the right the original hangars, passenger terminal, the old central tower, and the command museum are located.

THE CURTISS R5C COMMANDO. This transport aircraft could accommodate 40 fully equipped Marines or 33 litter patients.

THE FAIRCHILD R4Q PACKET OR C-119 (A.K.A. FLYING BOXCAR). This transport was first flown by the Air Force in 1947. It appeared in Marine Corps units in 1952. The C-119 was replaced by the C-130 in 1961.

THE DOUGLAS R5D SKYMASTER. This transport aircraft was a military version of the DC-4, but was diverted for military use during WW II.

THE LOCKHEED KC-130 HERCULES. This KC-130 is using a JATO assist to expedite his takeoff.

THE LOCKHEED KC-130. VMGR-352 celebrates returning to MCAS El Toro after a long flight away from home. VMGR-352s skull and crossbones flag is supported by an exuberant crewman.

AERIAL REFUELING. This Lockheed KC-130 refuels a CH-53E Super Stallion.

Five
A FEW FAMOUS MARINES

Fame in the Marine Corps comes in many forms. More often than not, fame is achieved for accomplishing positive objectives, while at other times, infamy attaches to those for misdeeds. There are hundreds, if not thousands, of truly famous Marines and only a few can be mentioned here. If one asks why a person was chosen for this publication instead of someone who may have done much greater things, the answer would be, "Because these are the ones whose photographs the El Toro library possesses."

There are 15 Marine Corps aviation Medal of Honor recipients and nearly 120 Marine Corps Aces. The Marine Corps has famous astronauts, helicopter pilots, and firsts in many aviation areas that could, by themselves, fill a novel. Marines, like Capt. Don Aldrich (who has 20 enemy planes to his credit), First Lt. John Bolt Jr. (an ace in both WW II and Korea), Marion Carl, Joe Foss, Robert Galer, Robert Hansen, James Swett, Wilbur Thomas, Ken Walsh, and many more deserve tribute beyond what can be done here. This chapter portrays a few good men of interest who once wore Marine green.

Pappy Boyington holds the honor of being the Marine Corps' most celebrated ace. Earning his wings in 1937, he flew at Quantico with Aircraft One for four years. Four months before the Japanese attacked Pearl Harbor, Boyington resigned from the Marine Corps and joined the Flying Tigers. Boyington became an ace prior to the U.S. entry into WW II by shooting down six Japanese planes flying with the famous Flying Tigers. Rejoining the Marines in 1942, he landed at Guadalcanal as the executive officer of VMF-121. While recovering after breaking his leg in an accident, he managed to form his own squadron (Black Sheep VMF-214) with rejects, misfits, and youthful inexperienced pilots who called him "pappy." During the squadron's first action, Boyington shot down 14 aircraft over a period of 32 days. In action over Rabaul, he was seen shooting down number 26 prior to disappearing in the fog of battle. News spread that Boyington had been shot down and captured by the Japanese, but not until he scored numbers 27 and 28. Boyington received the Medal of Honor upon his return from a Japanese POW camp.

> Why in hell can't the Army do it if the Marines can? They are all the same kind of men . . . why can't they be like Marines?
>
> —Gen. John J. Pershing U.S. Army

PAPPY BOYINGTON, COMMANDING OFFICER OF VMF-214.

PAPPY BOYINGTON.

TYRONE POWER. Tyrone Power was a famous movie star in the 1940s and 1950s. He refused a commission, but worked his way up to first lieutenant and earned his wings.

TYRONE POWER. In addition to being a famous movie star, he earned the reputation in the Marine Corps as being a "regular guy."

TYRONE POWER. When word leaked that he had done a little time in a Marine Corps Brig, there was no doubt that he was a real Marine.

JOHN GLENN. A Naval Academy graduate commissioned in 1943, Glenn flew 59 combat missions during WW II after he finished flight school. During the Korean War, he flew another 63 combat missions and shot down three MIG's during the last nine days of the war. He also set a transcontinental speed record in the F8U Crusader. John Glenn was the first American to orbit the earth. On February 20, 1962, he made three earth orbits for a total of 4 hours and 55 minutes. He retired from the Marine Corps as a colonel in 1964.

Senator John Glenn and the crew of the Space Shuttle Discovery join New York City Mayor Rudolph Guilliani in front of New York City Hall during the playing of the National Anthem at the end of their ticker tape parade Nov. 16. More than 500,000 people watched the parade through the "Canyon of Heroes."

Sgt. Jeffrey Castro

SEN. JOHN GLENN AND THE CREW OF THE SPACE SHUTTLE DISCOVERY. John Glenn's return to space in 1998 establishes him as the oldest person to travel in space.

CAPT. TED WILLIAMS. Commissioned as a second lieutenant in 1944, he was at Pearl Harbor in 1945 when WW II ended. Upon discharge in December 1945, he returned to his professional baseball career. Ted Williams received the professional baseball MVP award in 1949 while playing for the Boston Red Sox. In 1952, he was called back to duty in the Marine Corps to serve in Korea. Upon arrival in Korea, Ted served in the same squadron as John Glenn.

TED WILLIAMS IN PILOT TRAINING CLASSES. Capt. Marsh Austin shows Ted Williams a thing or two about nobs and switches in this photograph.

PILOT FORMATION ON THE GROUND. Ted Williams could be seen in a standard pilot formation when he was not flying. When not flying, pilots are often found standing in a circle on the hangar deck exchanging profundities.

Six
THE SIXTIES AND SEVENTIES

In 1963, El Toro celebrated its 20th anniversary; 75,000 people attended the annual El Toro air show that year. Fifteen thousand visitors attended the first air show on the nations first Armed Forces Day in 1950; the number would grow to two million visitors by the last airshow in 1997.

The Vietnam War incrementally turned into a full scale conflict during the 1960s, and the nation discovered for the third time in 20 years it was ill-prepared to go to war. El Toro once again became a debarkation point for servicemen and women embroiled in the toils of war. President Lyndon Johnson and later Richard Nixon landed at El Toro frequently to talk to the Marines, as well as conduct national business. The early 1940s marked a significant transition from slow, cumbersome bi-wing aircraft to the sleek, hi-performance propeller-driven fighters. The early 1950s marked a significant transition in aviation from propeller planes to jets. The '60s introduced similarly significant transitions where names like Phantom, Skyhawk, Intruder, Sea Stallion, Sea Horse, Sea Knight, Cobra, UH1, and Hercules changed the face of Marine Corps aviation like Shooting Star, Banshsee, Panther, Fury, Skyraider, Skynight, Corsair, Crusader, and Cougar. The aircraft of the '60s were more expensive, faster, noisier, and far more capable; they arrived just in time to serve in the Vietnam War. During the next ten years of war, personnel and squadrons rotated to and from El Toro and Vietnam in support of the First Marine Aircraft Wing.

At the end of the Vietnam War and the fall of Saigon, thousands of Vietnamese refugees relocated to the U.S. with little more than the clothes on their backs. During May of 1975, flights delivering Vietnamese and their families, who were loyal allies during the war, arrived at El Toro at the rate of one per hour. Over 50,000 people arrived as part of Operation New Arrival and then departed for processing locations across the U.S. "Freedom Villages," as they were called, amounted to tentcamps that oriented refugees into the American way of life.

With the cessation of the Vietnam War, defense budgets were slashed once again and parts of the armed forces languished. El Toro and Tustin lead the Marine Corps in the introduction and perfection of aerial weapons deployment and training, as well as the perfection of helicopter assault tactics and strategy. The 3d MAW developed the Weapons and Tactics School located in Yuma, Arizona, to teach advanced aviation war-fighting skills.

> The appearance of Marines on foreign soil has always indicated the beginning of an extremely dangerous military adventure.
> — Krasnaya Zrezda (Red Star) Moscow

EAST VIEW OF MCAS EL TORO. This photo was taken at 10,000 feet on July 10, 1962.

THE THIRD MARINE AIRWING BAND ON EL TORO'S PARADE DECK.

PASS AND REVIEW. During change of command ceremonies, squadrons march past the outgoing and incoming commanding officers.

EL TORO SEARCH AND RESCUE. This H-34 helicopter stands ready to respond to any emergency situation.

A SIKORSKY H-37 (HR2S). This H-37 is positioned at MCAS El Toro as backup for the El Toro Search and Rescue Squadron. Helicopters from LTA also support El Toro by carrying parts, personnel, or general cargo.

DANANG, SOUTH VIETNAM. This aerial view of Danang looks north on January 15, 1969.

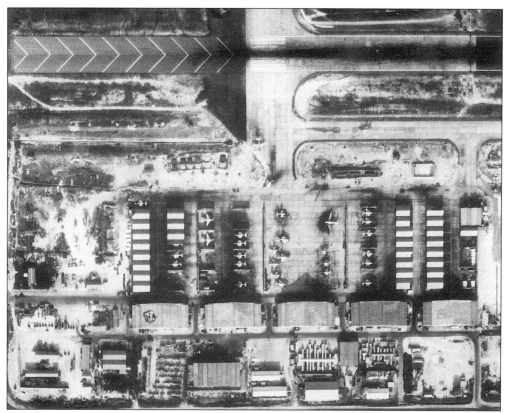

THE MARINE AIRCRAFT GROUP 11 AREA AT DANANG, SOUTH VIETNAM.

RETURNING FROM A COMBAT MISSION. This F-4 Phantom returns from a mission in Vietnam.

THE FIRST A4B LANDING. This A4B Skyhawk landed first at the recently completed airfield at Chu Lai, South Vietnam, on June 1, 1965.

HMM-161'S YANKEE ROMEO 31. This CH-46 arrives at a landing zone near the "Rockpile" in South Vietnam to bring combat supplies and evacuate casualties. (USMC Photograph.)

THE PRESIDENT. President Nixon landed at MCAS El Toro frequently during the Vietnam conflict when visiting his western White House in San Clemente, California.

THE MARINE AIRCRAFT GROUP 16 AREA AT MARBLE MOUNTAIN, RVN.

T-28 TRAINER. The T-28 replaced the SNJ Texan as the Marine Corps primary Trainer. This aircraft was used to train Navy and Marine Corps pilots for 30 years and was retired in 1984.

VMF-312. VMF-312 relocated to El Toro in 1962. Here, an impressive ordnance display shows the Crusader's power. In 1965, VMF-312 relocated to Japan. (USMC Photograph.)

A-4D-1 SKYHAWKS ON EL TORO'S FLIGHT LINE. (USMC Photograph.)

EL TORO STAFF NON-COMMISSIONED OFFICERS ACADEMY. Marines attending the academy are practicing close order drill by the landmark water towers.

CABINS IN THE MOUNTAINS. These cabins on Big Bear Mountain are part of El Toro's Morale Welfare and Recreation Department; they are available to Marines in need of R&R.

MARINE COLOR GUARD AND F4 PHANTOM.

Seven
BASE RENEWAL

The 1980s saw the introduction of the F/A-18 Hornet and AV-8 Harrier into the Marine Corps inventory. Both aircraft represent enormous leaps in tactics and strategy as well as another very significant transition in military design and employment. The Hornet has the capability to perform the missions previously flown by the A-6 Intruder, the F-4 Phantom, and the A-4 Skyhawk, while the Harrier redefines tactics along the FEBA with its vertical takeoff and landing capability.

President Reagan began the 1980s by picking up the gauntlet thrown down by the Evil Empire. The U.S. military that had languished badly in the 1970s was restored in the '80s to become the undisputed leader among military super powers. Across the board aircraft, equipment, facilities, training, and personnel benefited from Reagan's determination to defeat communism and the enemies of democracy. MCAS El Toro benefited enormously, resulting in a Model Installation Award.

Each year El Toro's annual airshow attracts thousands of daily viewers to the air station. The Blue Angels are always the spectacular main event that keeps everyone's attention. During the 1985 El Toro Airshow, an AT6 SNJ WW II Trainer aircraft crashed into the MCAS El Toro Chapel. The chapel was empty at the time, but unfortunately both the pilot and his passenger died in the accident. The chapel was completely destroyed with only a few artifacts surviving the fire. In March 1987, a groundbreaking ceremony was held for the new Chapel that served El Toro upon completion to base closure.

The mid-1980s saw the early remnants of the Marine Corps' only Aviation Museum assemble at El Toro. The museum has a total of 41 airframes ranging from WW II to a modern-day F/A-18 Hornet. Upon relocation to Miramar the museum will call itself the Flying Leatherneck Aviation Museum and embark upon a rebuilding plan that promises to result in one of the finest aviation museums in the nation.

Retreat hell, we just got here.
– Col. Frederic M. Wise at Belleau Wood, France

Come on you sons of bitches . . . do you want to live forever?
—Gunnery Sgt. Daniel Daly at Belleau Wood, France

MCAS El Toro Main Gate. The Model Installation banner is displayed proudly at El Toro's main gate.

THE NEW OFFICERS CLUB BAR. El Toro's new Officers Club bar has seen many notorious events and heard untellable war stories. The ever-dutiful "Rita" attends her duties behind the bar.

THE OFFICERS CLUB DINING AREA ADJACENT TO THE BAR. Photos and memorabilia adorn the walls offering Marines a pleasant place to spend their off-duty time.

THE NEW F-18 SIMULATOR BUILDING. These hi-tech, computer-operated, full video and motion simulators provide pilots with aviation skills before they step into the real thing.

FIRE STATION #1.

THE AFTERMATH OF AN AIRSHOW TRAGEDY. The chapel was completely destroyed when an airshow aircraft crashed; it was rebuilt a couple years later.

1987 EL TORO AIR SHOW. This aerial view shows the crowd gathering to watch the Blue Angels perform.

AERIAL REFUELING. This pilot has the aerial refueling basket stuck on his probe. This occurs occasionally when pilots don't use proper technique.

A F4 PHANTOM OVER EL TORO.

A4s and A6s on El Toro's Flight Line.

A6 Mechanics Working Feverishly.

A4 Mechanics Checking the Ejection Seat.

The A4 Flight Line.

An A6 Intruder on El Toro's Flight Line.

F4 Phantom Looking for Targets.

Two A4 Skyhawks displaying the Black Sheep logo.

Four A4 Skyhawks from VMA-311.

A FLOCK OF INTRUDERS WITH DIFFERENT TAIL FEATHERS. (Photograph Courtesy of Steve Dumovich.)

BAD WIND. Everyone stationed at MCAS El Toro knows about the Santa Ana winds. No one could believe the winds turned this CH-53 upside down. This model CH-53 weighs 13 tons empty.

THE VMFA-531 "GREY GHOSTS" LOGO. Activated in 1942, the Grey Ghosts saw action in WW II and Vietnam.

F/A-18 HORNET WITH BOMBS. The Hornet is a high performance tactical aircraft that can operate from aircraft carriers or land bases. It has a speed of 1.8 mach, a ceiling of 50,000 feet, and can carry more than 17,000 pounds.

F/A-18 Hornet with Missiles.

F/A-18s and KC-130 Hercules.

Four F/A-18s from Different Squadrons in Formation.

Eight
1990s
Desert Storm & BRAC

The 1990s yielded the fruits of President Reagan's efforts with the unexpected collapse of the Evil Empire and proof of America's military might by the near total destruction of the world's fourth largest military machine, Iraq. During Desert Storm, President Bush brought together a coalition of Christian, Muslim, and Jewish forces to defeat Iraq in what historians will undoubtedly remember as one of history's best executed political and military operations. When Iraq invaded Kuwait, the Marine Corps was ready. When President Bush deployed the Marine Corps in support of Desert Storm the 3d MAW became the largest airwing in Marine Corp history operating units from bases in Shikh Isa, Bahrain, Jubail, Tanajib, Al Khanjar, Al Mishap, King Abdul Aziz Naval Base, and Minifa Bay, Saudi Arabia. The expeditionary airfield at Al Khanjar became the largest in Marine Corps history. In typical Marine Corps fashion, the 3d MAW deployed to Southwest Asia in only 25 days and was the first large aviation unit to be combat capable in the theater.

Upon the conclusion of Desert Storm, El Toro Marines sailing home on Navy ships were diverted to the Philippines to assist the post volcano humanitarian mission, and Bangladesh where post monsoon floods caused a national disaster. Not long after arriving home, El Toro and Tustin Marines once again found themselves redeployed "overthere" to aid in the Somalia humanitarian effort.

With the Evil Empire vanquished and no super powers poised to attack, "BRAC" (Base Realignment and Closure) has eliminated more U.S. military bases than any enemy could have ever dreamed of. Many of these base closures were necessary in the post-Cold War world. Many of these base closures have been controversial because they appear to be eliminating a military war-fighting asset with no replacement or substitute. MCAS El Toro and MCAS Tustin fell under the BRAC attack.

One person with a belief is equal to a force of 99 who have only interest.

—John Stuart Mill

F/A-18 Hornet Stalking a Russian Bomber.

OIL WELLS BURNING DURING DESERT STORM.

THE MWSS (MARINE WING SUPPORT SQUADRON) 472 DET C LOGO. This little known and unheralded reserve squadron activated for Desert Storm arrived at El Toro in December of 1990. Nearly 500 Marine Reserves were to assume MWSS duties for the two active duty squadrons deployed for Desert Storm. They also augmented the Yuma and Camp Pendleton Air Stations and the Twenty-Nine Palms MWSS.

F/A-18 Hornet on the Roll with the El Toro Hills in the Background.

A Hornet Swarm Beneath a KC-130 Hercules.

A LANDMARK FALLS. The landmark El Toro water towers provided visual cues for El Toro pilots for five decades. Their destruction is an ominous indication of things to come.

EL TORO BOULEVARD. Looking at the boulevard from the intersection of Marine Way reveals a well-kept, mature, and even elegant base that will be missed by all who served here.

THE EL TORO HEADQUARTERS BUILDING ENTRANCE. If trees could talk, a few in this picture could tell the whole story.

THE MCAS EL TORO COMMAND MUSEUM (A.K.A. JAY W. HUBBARD AVIATION MUSEUM). This is the Marine Corps' only aviation museum and will reopen at MCAS Miramar upon relocation. At Miramar, the museum will be called the Flying Leatherneck Aviation Museum.

FRONT VIEW OF THE EL TORO COMMAND MUSEUM. In the late 1980s, a small group of dedicated volunteers came together to form the Marine Corps' first and only aviation museum. With the support of successive commanding generals the El Toro Museum became an official Command Museum in 1993. It was named after BGEN Jay W. Hubbard (USMC), who led the charge in creating the museum.

SNOW CAPPED SADDLEBACK MOUNTAIN. This view from the museum shows an A4-F Skyhawk in the foreground of the beautiful snow capped Saddleback Mountain.

BRIGADIER GENERAL ROBERT MAGNUS. Assuming command of Marine Corps Air Bases Western Area and MCAS El Toro on June 26, 1997, Maj. Gen. Magnus reported from Headquarters, U.S. Marine Corps, where he served as Head of Aviation Plans, Programs, Doctrine, Joint Matters, and Budget Branch. He subsequently served as assistant deputy chief of staff for Aviation until being assigned to MCAS El Toro. Now a major general, he relocated his headquarters to MCAS Miramar, California, but holds the distinction of being the last commanding general of MCAS El Toro. Maj. Gen. Magnus had the colossal multi-billion dollar responsibility of closing El Toro, transferring all El Toro and MCAS Tustin squadrons and assets while overhauling MCAS Miramar to meet Marine Corps needs. He did this with a clarity of vision and charisma that eliminated much of the difficulty and friction normally associated with a task of this proportion.

LAST CHANGE OF COMMAND. Maj. Gen. Robert Magnus, commander, Marine Corps Air Bases Western Area, shakes hands with Col. Stephen F. Mugg, the new and last commanding officer of MCAS El Toro.

THE FOURTH MARINE AIRCRAFT WING LOGO. The Marine Corps Reserve has had a strong presence at El Toro since WW II. Modernly Fourth MAW's MAG-46 moved to El Toro from Los Alamitos in 1971 while MAG-46 Det B was formed in 1996 when HMH-769 relocated to El Toro from NAS Alameda, California.

HMM-764 Moonlighters. HMM-764 is a reserve unit flying CH-46s that has been stationed at El Toro since the mid-1970s when it relocated from MCAS Tustin, California.

VMA-134 Smoke. When VMA-134 was reactivated in 1958, it flew the A4 Skyhawk as a reserve unit at Los Alamitos, California. It later transitioned to MCAS El Toro, as well as to the F4 Phantom. In 1989, the squadron became the first Marine reserve squadron to transition to the F/A-18 Hornet, highlighting the modernization of the Marine Corps reserves.

HMH-769 ROADHOGS. The Roadhogs arrived at El Toro in 1996 from NAS Alameda, California, flying the Sikorsky RH-53D Sea Stallion. HMH-769 was the only reserve helicopter transport squadron to serve in the combat theater in Desert Storm. HMH-769 transitioned to the CH-53E Super Stallion in 1996. (Photograph by Rick Mullen.)

Dirge By A Marine Corps Chopper Pilot

by Quinn Mulhearn

On a standard day by early morning's light
readying my craft for a long day's flight,
excited anticipating the sound of might
and confident everything is right,

Lifting vertically from earth's beckoning grasp
hugging the surface with an exhilarated gasp
roaring skyward with wings swept round,
slipping forward napping the ground
and other thing you would not dare in this surround,

Beneath the clouds and over mountain top
into canyons deep stealing ethereal things powerless to stop
spiraled and spun and stopped in the meadows sleep
I've seen the eagles face, the hawk and the ravens sleek,
and too the gulls below, while hovering low and slow
and followed the twisted stream to waterfall where owl and osprey go,

Guiding my trembling craft through rain and snow and fog on seashores edge
I've hung and flung and dipped in turbulent wind into earth's deepest wedge
throughout the day and into the night in ways you could never know,
I've earned my wings by lunar light
skimming the earth in death defying flight
fearless as a winged seraph shielded by God's halo,

And, in the bonded company of aircrew and natures grace revering vista's grand
I too in silent vigil stand
not anxious to meet God in the fury of battles plan
but if I do I'll shake his hand
and thank him for the gift of flight

If for reasons grand or upon a whim
God diverts my route of flight
and instead sends me to the eternal night
I will forever be grateful for having had the opportunity to fight
and will always treasure the thrill of flight.
and hope he forgives my mortal sin

Upon arrival at Satan's gate
I'll step up and take charge of my fate
announcing name, rank, and numbers I'll be praying
that Satan will send me to his airwing flying.

Nine
MCAS Tustin (LTA)

Construction began on April 1, 1942, to construct hangars and tie-downs for a Lighter-than-Air (LTA) facility in the middle of the second bean field purchase from the Irvine Ranch. On October 1, 1942, the Santa Ana Naval Air Station (LTA) was commissioned by its first C.O. Capt. H.N. Coulter USN. Each of the two hangars is 183 feet tall or the equivalent of 18 stories and 1,026 feet long. The exterior width at 297 feet yielded a 241,110-square-foot hangar for the storage of seven full-sized blimps and eventually Marine Corps helicopters. In 1947, the Naval Air Station was officially decommissioned, but on May 1, 1951, with the advent of the Korean War, the base was reopened by the Marine Corps as Marine Corps Air Facility Santa Ana, California. The first Marine Corps commanding officer was Col. L.H. McCulley while Maj. R.H. McKenzie Jr. was the last commanding officer. Initially, 19 Marine officers and 243 enlisted personnel were assigned to the facility with only one helicopter squadron (HMR-161) and one observation squadron (VMO-2). However, by 1952, two Marine Air Groups (MAG-16 and 36) and 13 squadrons arrived. When the Marine Corps recommissioned the base, only 200,000 people lived in Orange County.

LTA has always been considered the "pearl" of Marine Corps bases. In addition to its perfect location among the southern California orange groves and proximity to all of the southern California amenities, it also provides a nearly perfect environment for training helicopter pilots. The nearby Saddleback Mountains and adjacent foothills have 13 confined area mountain landing sites that every "PUI" (puuee) (Pilot Under Instruction) training at LTA has learned to hate and love. The base is also only a few miles from the Pacific coast and provides training opportunities for operations aboard Navy ships. To the northeast the Marine Corps Expeditionary Airfield at Twenty-nine Palms provides pilots all the challenges of flight in a hot, heavy, turbulent, and high-density altitude environment. Not far away to the southeast the Marine Corps base at Yuma, Arizona, doubles if not triples the opportunity to experience desert flight. When helo pilots tire of desert flight, all they have to do is fly up the I-395 highway to the Marine Corps winter training facility at Bridgeport, California. Here, landings between 8,000 and 10,000 feet in confined area zones covered with 6 feet of snow and surrounded by 150-foot pine trees brings out the best in every helicopter pilot.

On September 1, 1969, the base was redesignated Marine Corps Air Station (Helicopter) Santa Ana, California. On April 23, 1976, it was annexed by the City of Tustin; on June 1, 1978, the name was changed to MCAS (H) Tustin. After BRAC and downsizing, LTA became a Marine Corps Air Facility a second time in 1998. The base was closed a final time in 1999.

> The Marine Corps is the Navy's police force and as long as I am President that is what it will remain. They have a propaganda machine that is almost equal to Stalin's.
>
> —President Harry S. Truman

THE LAST CO. Colonel S.F. Mugg was MCAS El Toro's last commanding officer. He made this statement:

Fellow Marines and Friends of Marine Corps Air Stations El Toro and Tustin: As we come to the closing moments of these two historic bases, we reflect on the dramatic impact that they have had on the County, State and the Nation. Both were born of necessity from the bean fields of the Irvine Ranch at the beginning of our Nation's global struggle against the forces of fascism. Both were instrumental in the long struggle against communism as they housed and trained the forces of democracy for combat in Korea, Vietnam and the eventual destruction of the Berlin Wall signaling the end of the 'Cold War.'

Both have served through one of the most troubled and conflicted periods of the world's history and played their parts in the establishment of the United States as the preeminent world power at the close of this millennium. Both served as economic engines that transformed the rural agrarian Orange County into one of California's most important and successful urban industrial areas. For more than 50 years, they provided the gateway through which hundreds of thousands of America's finest citizens moved enroute to serve their Nation on foreign shores and it was that river of people that truly made these bases special.

It was the sacrifice, the dedication and endless efforts of all who have served aboard El Toro and Tustin that has made the greatest impact. We became neighbors, trained for combat, built communities, deployed in the defense of our Nation and returned here to live after our military service was complete. Our mark on the County can be seen in every corner as Marines and Sailors helped transform the landscape. It comes down to the people who have served here and as long as we remember them, we will preserve the Espirit-de-Corps that El Toro and Tustin symbolize. So, as you enjoy these photographs, please stop for a moment and listen for the 'sound of freedom' that has thundered through the Saddleback Valley for 56 years. It's still there, echoing in the canyons and over the shoreline, an indelible memory for all who have shared in the history of Marine Corps Air Stations El Toro and Tustin.

The MCAS Tustin Logo.

Early Hangar Construction at LTA.

RAPID PROGRESS. The enormous hangars were completed surprisingly quickly.

COMMISSIONING CEREMONY AT LTA ON OCTOBER 1, 1942.

AERIAL VIEW OF LTA LOOKING FROM THE MAIN GATE TO THE HEADQUARTERS BUILDING.

BLIMP HANGAR. Seven blimps fit comfortably in each hangar.

EARLY AERIAL VIEW OF LTA. The runway splits the two large hangars while the circular landing pads were used for independent operations for both blimps and helicopters.

TOWING A BLIMP OUT OF THE HANGAR PREPARING FOR FLIGHT.

COL. H.C. FREULER. Freuler replaced Colonel McCulley as commanding officer.

EARLY AERIAL PHOTOGRAPH OF LTA AFTER THE MARINES LANDED.

HEADQUARTERS. Building #4 served as the headquarters building for the base commanding officer and the CO of H&HS.

STATION CHAPEL.

THE FIRST HELICOPTERS. H-19s were the first Marine Corps assault support helicopters. HMR-161, later HMM-161, was the first Marine Corps assault support helicopter squadron and was deployed to the Korean War shortly after being formed.

THE H-19s FRONT MOUNTED RECIPROCATING ENGINE.

THE SIKORSKY H-37 (HR2S). The H-37 satisfied many of the requirements for a heavy lift helicopter.

THE SIKORSKY H-37 WITH OPEN CARGO DOORS. This original front load design was moved to the rear of H-53 helicopters for many very technical reasons. One not very technical reason was the impact of birds crashing through the cargo door windows.

THE SIKORSKY H-34 (HUS) INSIDE A LTA HANGAR.

THE LTA MAIN GATE IN THE 1960s.

THE STAFF NON-COMMISSIONED OFFICERS CLUB. This club, conveniently located next to the Hangar, hosted many "bosses nights," as well as "other" forms of entertainment.

UNUSUAL ATTITUDES. Ground personnel are giving this CH-46 a wide berth.

COLOR GUARD. The MCAS Tustin Color Guard stands proudly in front of a hangar.

Two CH-46s from HMMT-302.

A CH-34 (HUS) Sikorsky Helicopter from HMR-363.

A CH-53 with Blades Folded.

HELICOPTER AIR CONDITIONING. Opening the forward avionics compartment door on a hot summer day provides a welcome breeze. The trick is to close the door before takeoff.

FOUR HELICOPTER SQUADRONS OCCUPY ONE HANGAR AT LTA.

A Mechanic Checking the Tail Rotor. This photo shows the relative size of a CH-53E tail rotor.

A CH-53E Ferrying Marines in a Non-Tactical Movement at Twenty-nine Palms. The CH-53E can carry 55 fully equipped combat Marines. (Courtesy of Sikorsky Aircraft.)

A CH-53E SUPER STALLION. The Super Stallion is the largest helicopter outside Russia and the highest performance helo in the world. At 99.5 feet in length and at a maximum gross weight of 73,500 pounds, the Super Stallion has a mission payload capability of 32,000 pounds. (Courtesy of Sikorsky Aircraft.)

CH-53E Super Stallions Waiting for Fuel. (Courtesy of Sikorsky Aircraft.)

A CH-53E Carrying an Artillery Piece. (Courtesy of Sikorsky Aircraft.)

ANOTHER FIRST. First Lt. Sarah Deal is the first woman Marine pilot.

THE LAST MCAS TUSTIN COMMANDING OFFICER. Col. Tom Caughlin was the last commanding officer of LTA as a Marine Corps Air Station.

The Last CO of LTA. Maj. Richard H. McKenzie Jr. served as the very last commanding officer of LTA as a Marine Corps Air Facility.

The Last Sunset for MCAS El Toro and MCAS Tustin.